Empress Adelaide Dances at 8

A time-telling tale

Story by M.W. Penn
Photographs by Stephanie Anestis

With an activity by
Adam Goldberg, Ed.D.

Story © 2012 M.W. Penn
Photographs © 2012 Stephanie Anestis

ISBN 978-0-9840425-1-7
Library of Congress Control Number: 2012949227

Printed in the USA on acid-free paper that contains no material from old-growth forests, using ink that is safe for children.

Empress Adelaide Dances at 8 is published by MathWord Press.

All rights reserved. No part of this book may be reproduced in any form or by any means, electronic or mechanical, including photocopying and recording, or by any information storage and retrieval system, without permission in writing from the publisher.

Empress Adelaide wakens at 6
And eats oatmeal
And orange peel
On bread.

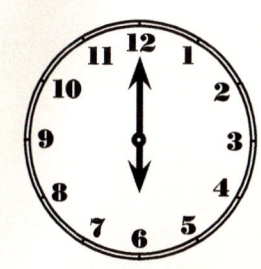

At 7, sweet Adelaide dresses for school

In a gown,
With a crown
On her head.

Empress Adelaide and her chauffeur
   Drive to school,
      As a rule,
         Before 8.

Empress Adelaide meets with the friends

From her class
On the grass
Near the gate.

Adelaide studies from 9 until noon,

Learns of nations  *crustaceans*
quotations *dictations*

an empress, as you'd guess, must be smart.

summations  translations and art.

In a tent
That is sent
By her staff.

After their lunch until class starts at 1,

The friends run
In the sun
And they laugh.

2 O'clock: dinosaurs, puzzles, and crafts
Or a book, in a nook, occupy
Empress Adelaide and all her friends.

All too soon
Afternoon
Passes by.

 Homework and studies are done until 5,

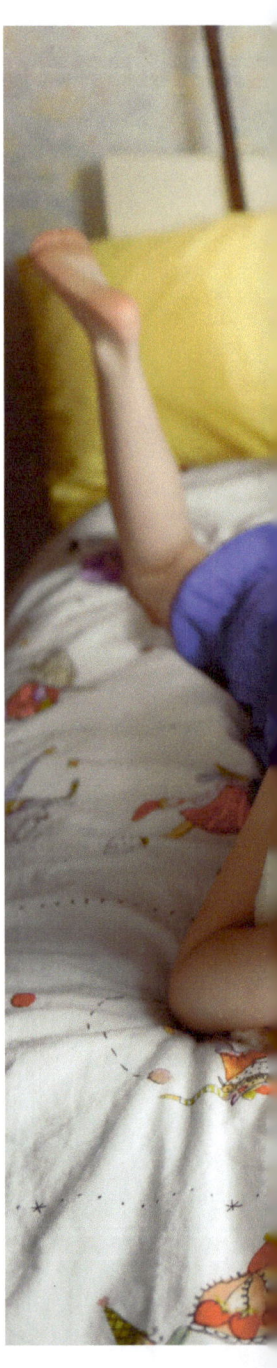

Then she rests

And gets dressed
For the ball.

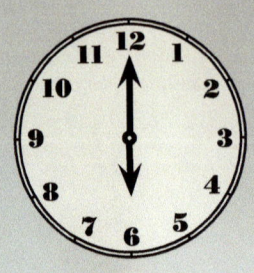

Adelaide's guests arrive promptly at 6.
They all dine
On the finest
Buffet.

Soon after dinner they glide to the hall,

As the grand
Palace band
Starts to play...

Empress Adelaide
    dances at 8,

With the Earl
With a curl
    In his hair.

Soft moonbeams

And sweet dreams

Fill the night.

Rodeo Adelaide wakens at 6,
And eats oatmeal,
And orange peel,
On bread.

A time-telling fairy tale activity

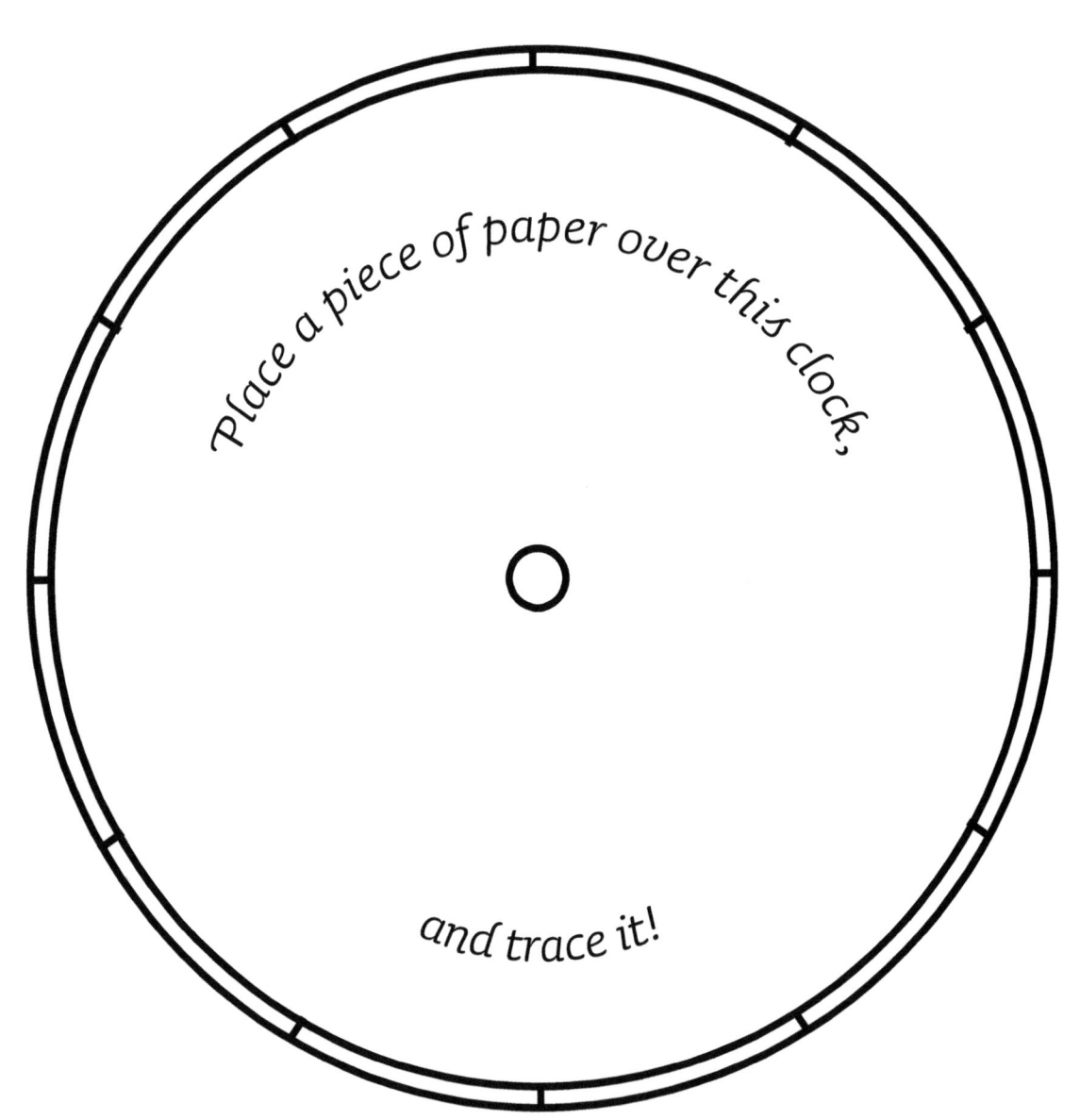

Place a piece of paper over this clock, and trace it!

Cut out the clock you've traced, and write the hours on the clock face.

Trace these hands on a sheet of paper and cut them out too.

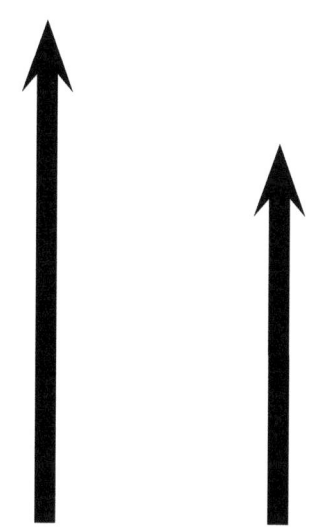

On your new clock, place the ends of both hands on the dot at the center of the clock with the arrows pointing to any hour of the day. Then decide what you and Adelaide would do at that time and imagine a story about it.

Your story might include the children from Adelaide's school, guests at her dinner or friends from other stories that you enjoy. Magic carpets, purple monkeys and anything else you can imagine are allowed.

Have a great time!

About the author

M. W. Penn is an award-winning author of 15 children's books focused on mathematics and an activity book for early elementary mathematics based on the Common Core State Standards. Her poetry appears in *Highlights for Children* magazine and several anthologies. She presents sessions in interdisciplinary literature at NCTM and NCTE conferences across the country. Visit her website at **www.mwpenn.com**.

About the photographer

Stephanie Anestis is a photographer specializing in environmental portraits of children and families. She loves documenting the lives of children by capturing moments of spontaneous emotion and expression. Her work has appeared in juried shows and art spaces throughout Connecticut. View her work at **www.stephanieanestis.com** and keep up with her current projects on her blog, **blog.stephanieanestis.com**.

About the math activity creator

Adam Goldberg is an education professor at Southern Connecticut State University, where he specializes in teaching mathematics to elementary teachers. He lives in Connecticut with his wife Kerri and daughter Isobel.

Dedications

M.W. Penn would like to dedicate this book to Emily, who never wanted to be an empress, and always made us smile.

Stephanie Anestis would like to dedicate this book to her husband, who convinces me that dreams are within reach.

www.ingramcontent.com/pod-product-compliance
Lightning Source LLC
Chambersburg PA
CBHW042126040426
42450CB00002B/93